How Does It Grow?

# HOW DOES GRASS GROW?

By Kathleen Connors

Gareth Stevens PUBLISHING

**Please visit our website, www.garethstevens.com. For a free color catalog of all our high-quality books, call toll free 1-800-542-2595 or fax 1-877-542-2596.**

**Library of Congress Cataloging-in-Publication Data**
Names: Connors, Kathleen, author.
Title: How does grass grow? / Kathleen Connors.
Description: New York : Gareth Stevens Publishing, [2022] | Series: How does it grow? | Includes index.
Identifiers: LCCN 2020012337 | ISBN 9781538263150 (library binding) | ISBN 9781538263136 (paperback) | ISBN 9781538263143 (6 Pack) | ISBN 9781538263167 (ebook)
Subjects: LCSH: Grasses–Juvenile literature.
Classification: LCC SB197 .C65 2022 | DDC 633.2–dc23
LC record available at https://lccn.loc.gov/2020012337

First Edition

Published in 2022 by
**Gareth Stevens Publishing**
111 East 14th Street, Suite 349
New York, NY 10003



Editor: Kristen Nelson
Designer: Katelyn E. Reynolds

Photo credits: Cover, p.1 Tim Hawley / Photographer's Choice / Getty Images Plus; p. 5 Ning Li/Moment/Getty Images; pp. 7, 24 (seeds) georgeclerk/E+/Getty Images; pp. 9, 24 (soil) Faba-Photograhpy/Moment/Getty Images; p. 11 oxign/E+/Getty Images; pp. 13, 24 (roots) redmal/E+/Getty Images; p. 15 VR19/ iStock/Getty Images Plus; p. 17 Jacky Parker Photography/Moment/Getty Images; p. 19 Mint Images RF/Getty Images; p. 21 Sheryl Watson/ iStock / Getty Images Plus; p. 23 WILLIAM WEST/AFP via Getty Images.



Printed in the United States of America



CPSIA compliance information: Batch #CSGS22: For further information contact Gareth Stevens, New York, New York at 1-800-542-2595.

# Contents

Grass is all around us!
How does it grow?

Grass grows from seeds.
The seeds are tiny!

They grow in soil.

They need water.
They need sunlight too.

The seeds sprout.
Leaves and roots grow.

This can take about two weeks.

There are many kinds of grass.

Wheat is a grass.
People can eat it!

Ryegrass is green.
It may grow in your yard!

Cows eat grass.
It helps them grow too.

# Words to Know

roots

seeds

soil

# Index